The Rise And Fall Of Singer Manufacturing in Britain

End of Empire
By
Alex Askaroff

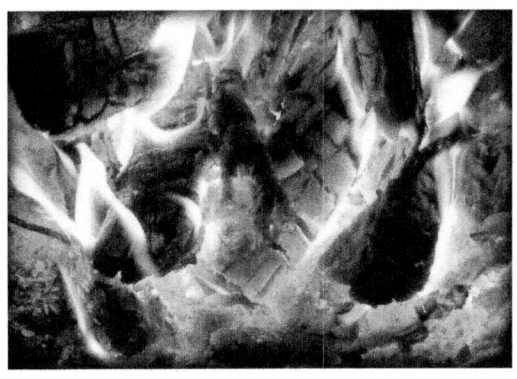

The rights of Alex Askaroff as author
of this work have been asserted by him
in accordance with the Copyright,
Designs and Patents Act 1993.

To see Alex Askaroff's work
Visit Amazon

The Rise And Fall Of Singer Manufacturing in Britain

This is no masterpiece. It's more a self-published labour of love from someone who has spent a lifetime in the sewing trade and a million hours gathering facts for you. From my vast profits of around sixpence a book I'll take up basket weaving. Why write it? Well no one else bothers! Please forgive my spelling, United Kingdom English, and enjoy it in the spirit it was written. Also I found it impossible to split into chapters. Just go with the flow.

Introduction
The Seeds Of Industry

The seeds of the Industrial Revolution had been growing slowly since Medieval Times here in Britain. When they did finally take hold, the changes were profound and dramatic. In a few years the kingdom went from horse and cart, hardly changed since the invention of the wheel, to steel and iron, smoke and furnace; from working the land to mills pricking the sky like an over-filled pin cushion. Sprawling cities grew to house the ever eager migration from the countryside.

Great names would soon be registered into the industrial history books of the world, from John Kay and his flying shuttle to Eli Whitney and his cotton gin. Industrial giants rose from the earth like Titus Salt. Not only did he build great factories to employ thousands, he also built the very towns they lived in.

None of the changes were more intense than in the small village of Kilbowie in Scotland. It transformed from wild open land into one of the mightiest factories on Planet Earth.

The cotton mills had perfected the mass production of fabrics. A hungry population needed a method to join that fabric quicker than by hand. A handmade dress could take weeks of precious time, and when people were working 70 hours, six days a week, that was hardly possible. The result was that most of the population, except the wealthy ruling elite, looked similarly drab, in mundane working clothes, with perhaps a change for Sunday church.

All that changed in 1851 when Isaac Singer brought onto the market the first practical sewing machine in history. He had combined all the ideas that had gone before to make a machine that actually worked. The result was explosive. By the 1870's the humble sewing machine had become the most wanted item on the planet. Why? Because with it, anyone could pop down their local market on Saturday and wear a new dress to church on Sunday! Cheap material and the sewing machine meant that beautiful clothes were no longer the sole domain of the upper classes.

For the first time in human history, personal expression in everyday clothing was being demonstrated by the working class. The effect was as dramatic as a meteorite hit. Beautiful women in beautiful clothes paraded along promenades and parks, pestered by elegant men in figure hugging suits. Poor migrant workers, pouring into the cities from the countryside could hardly believe their eyes. It was as if they had landed on an alien planet.

How could it be possible? It was possible because behind the scenes great factories had grown to supply sewing machines to meet that insatiable demand to join fabric. The biggest and best factory was Kilbowie at Clydebank, a vast and magical monster consuming raw steel and wood at one end and churning out mechanical marvels at the other.

The simple ability of joining cloth quickly has changed our world in the most dramatic style imaginable. You only have to look at the high street and Internet to see its profound effect on our daily lives. Every stitch of clothing we wear, every seat we sit on and bed we sleep in has some form of weaving and stitching. Amazingly the first complex mass produced product in the household is still going strong in our modern world. I mean, we may all be able to cope without our phones, but without our clothes!

Let's look at how Singer came to Britain and the spectacular rise and fall of Singer sewing machines.

Chapter One

Forging metal has been a skill shared by peoples and cultures across the world for millennia. In 1850s America, as the fires of civil war rumbled ever closer, Isaac Singer and his business partner, Edward Clark had taken that skill and turned it into an art form. They went on to perfect the forging of metal to bring us the first exceptional sewing machine in the world, the Singer Model A. They also used cutting edge technologies to build factories, marching ever forward to become the largest multinational sewing machine company the world had ever seen.

Isaac Singer

By 1865, as the final weapons of war in America fell silent, Isaac Singer and Edward Clark were ushering in mechanization on a global scale. They had also done it again, designing and building another amazing sewing machine. It would become a universal success, the fabulous Singer New Family Model 12. It would change the lives of millions and usher in a new era of mass production.

Edward Clark

The factories needed to build this beauty would spread, firstly across America and then Europe, starting in Mott Street, New York then expanding to Elizabethport, New Jersey in 1872.

The new factories included all the infrastructure, railways and docks, warehousing and container ships. This helped to kick-start the rebuilding of a shattered country, employing thousands, and pushing America onward to eventually become the world's first super power.

Singer factory Elizabethport, New Jersey. All the Singer factories were on large rivers for transport.

Singer would go on to invest millions (billions today), building factories around the world from Cairo, Illinois, to St. Petersburg, Russia, Monza in Italy, Wittenberg in Germany, Bonnieres on the river Seine (not far from Paris) France, Brazil and many more. The biggest of them all was at Kilbowie in Scotland.

The special machines that built the sewing machines (shipped around the world from Elizabethport) were so precise that for the first time, parts from any factory could be shipped to another and would integrate perfectly with the sewing machines that were being produced there.

It was the first true mass production on a global scale (forty years before Henry Ford had taken his first steps to make his Model T).

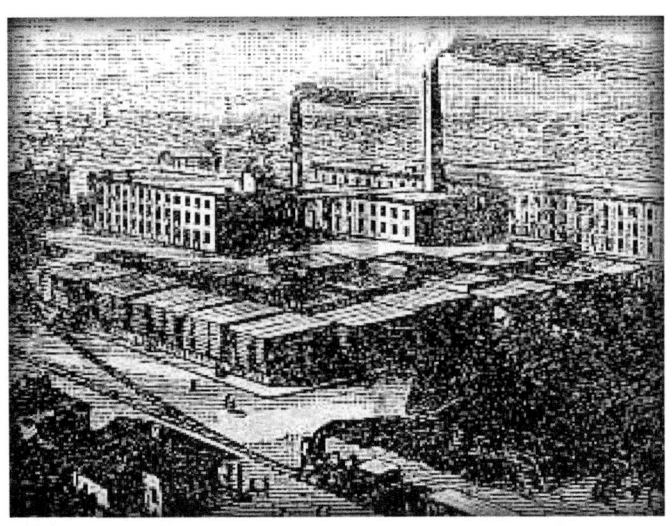

The Singer factory, South Bend Indiana

The Kilbowie plant, built in the cold Scottish hills, was to change a rural valley on the side of the River Clyde, into one of the greatest powerhouses of industry on Earth, employing at its peak (just before the outbreak of WW1) an estimated 14,000 workers. It would also bring prosperity to the lives of countless other families.

Mighty Kilbowie spread along the curving banks of the Forth and Clyde Canal. It became the pulsating heart of the Singer Empire. Raw pig iron went in one end of the 46 acre site and precision built engineering masterpieces poured out of the other. Eventually, by sea and rail, plane and road, they travelled across the globe to eager customers waiting with open arms. The factory lasted until Singers final death knell rang through the Scottish hills.

Chapter Two
How did it all start?

The Singer Company had been on a boom after Edward Clark (in the 1850's) had cleverly devised the first ever official hire purchase, lay-away, part payment, or instalment plan, allowing just about anybody that wanted a machine to buy a Singer.

The call for their machines (that cost almost a year's wage) was growing at an astronomical rate. Keeping up with this demand, from all four corners of the world, was a major problem for the first ever multinational company.

The first time that the name Singer was put across the front of the machine was on the new models coming out of Kilbowie after 1883.

The original idea of factories in Europe was simply to overcome the large costs of shipping heavy cast iron machines from America and to get around any import levies or duties imposed by the European powers. The other reason Singer chose Scotland was to get around the patents that were in force in England at the time.

In 1866, Edward Clark's cousin had been sent on a European tour to seek out areas for new factories. He reported back to Clark, who narrowed down the ideal locations and started to build small units. In 1867, plans by Clark were afoot to send George McKenzie (who was general manager of The Singer Sewing Machine Company and later president) to Scotland, to prepare the way for one enormous factory, instead of their smaller ones.

In the 1860's, 70's and 80's Singer was running several smaller sites in Scotland, mainly based around Glasgow. There was a site at Love Loan that was building complete machines for European distribution (from boxed parts shipped over from America).

There was the Bridgeton Site (opened in 1871) which actually made complete industrial models, plus the model 12 and 13 'New family'. There was the foundry at Bonnybridge some 20 miles away, and another at Govan Street in Glasgow.

The Kilbowie site had great potential, and as McKenzie was of Scottish descent, he was perfect for the job. The plan was simple: integrate all these smaller plants into one perfect sewing machine manufactory, capable of expanding with the market.

The last meeting at Oldway of Isaac Merritt Singer, George Ross McKenzie and Inslee Hopper. Isaac died shortly after this photo in the summer of 1875.

The factory needed as many natural resources as possible and Kilbowie offered these in abundance. Coal, steel, and seemingly endless forests for the cabinet makers to craft the wooden shells of the sewing machines. At Kilbowie craftsmen would perfect the steam bending of wood to create the first 'domed' lids for their sewing machines.

Then there was the mighty River Clyde, where huge ships could carry Singer's stock across the world. Access to railways from the city of Glasgow and major road links to the rest of the country were other bonuses.

An explosion in growth was to happen to the tiny rural hamlet after the first sod was lifted with a silver spade by George McKenzie in the spring of 1882. The factory would be powered by the latest 'cutting edge' steam technology. Puffing chimneys would replace the tall pines and steel rails would bring screeching trains shuddering in 24 hours a day.

780 automated machines would cut and make the screws, nuts and bolts, 180 more would make just the needles. Huge tubs of whale oil would quench and harden the hot steel shafts. Every facet of production would be handled in one massive plant with 57 separate departments and over two million square feet of factory floor. Some workers found themselves walking up to 20 miles a day around the gargantuan plant.

The massive Kilbowie Singer Factory on the banks of the River Clyde. In its prime it was the largest factory on Earth producing millions of technically perfect sewing machines. So perfect in fact that many are still sewing today. The Singer Company set the trend for mass production all over the world.

At one point Singer was the most well-known name on the planet.

Men hand finishing the Singer New family model 12. Before mass production all sewing machines were hand built and hand finished, what was referred to as 'file to fit'. Mass production eliminated the need for each machine to be individually built and subsequently reduced final costs to the public. Due to mass production Singer sewing machine prices dropped steadily for nearly 30 years while sales boomed.

As the Singer model 12 went into production, a population boom was to follow as farmer's sons left the hard toil of the land for the heavy thumping of the anvils, and the pouring of molten iron. The Singer models, 13, 15, and 17, soon followed, all for the first time with the Singer name emblazoned across the arm of the machine. The Singer model 15 went on to become the bestselling (and copied) sewing machine of all time.

Chapter Three
Mother of Pearl Singer sewing machines

End of shift at the Singer factory in Kilbowie. 14,000 people walked by the enormous clock tower every working day. Legend tells that the largest clock in the world was built so that no one had an excuse to be late in Kilbowie! It could be seen from every window for miles.

The main contract for building the plant was won by McAlpine's. It was their first major contract. Robert McAlpine soon split with his partner (Richmond) and took the contract on alone, using over 20 million bricks from his own brickworks. McAlpine later built his company into Britain's largest road builders.

The Kilbowie plant would not only have the latest machinery inside but was protected by fireproof walls of brick, iron and steel, incorporating a full sprinkler system. These innovative ideas saved the factory from any major fire damage.

By late 1883 the monster was fully breathing fire from huge melting and smelting plants as the pig iron was hauled in to be smelted.

The factory slowly evolved rather than opened with a bang. It didn't have the usual pomp and circumstance ceremony. A sleeping giant noisily rose from the Scottish soil and kept on growing for decades. Although production started slowly more trains were soon organised to bring fresh hands and materials to the plant.

Adverts eventually boasted that over one million complete machine every week were being made at Kilbowie, by a highly trained and skilled workforce unparalleled on the planet. It had become the largest factory the world had ever seen.

During this first dynamic period of manufacture, Singer produced something amazing, hand decorated mother of pearl sewing machines. Nearly every MOP Singer that has come to light (they are rare) have the Kilbowie early 1880's serial numbers.

Speculation is that there were a handful of these stunning machines made for local dignitaries and important visitors of the time, to show just how special the Kilbowie plant was. They were possibly all made by one individual, a specialist in pearl inlay and decoration.

Chapter Four
End of Empire

In a time of depravation and mistreatment of workers in Britain by the ruling elite, Kilbowie broke the mould. While Singer paid an average wage, the way they looked after their huge staff was far from the norm of the day.

They had their own cooks and staff, their own playing fields and sports teams, clubs and entertainment halls. All-in-all a marvel beyond all description to keep their workforce content which in turn allowed Singer to feed the ever-hungry demand for the humble sewing machine.

A strike by the women in 1911 (before the suffragettes were in their embryo stage) brought benefits like social clubs, dance halls and even beauty pageants. Their courage changed the workforce for the better, though equal pay for women was still just a dream.

Unfortunately, Isaac Singer (the little runaway who had lived on his wits, scraping a living from the dirt) was never to see his mighty dream being built in the north lands of Britain. Isaac died in July of 1875 at his opulent palace 'Oldway' in Paignton. However his dream did not die with him, eventually leading to one of the mightiest factories on earth.

One of the final images of Isaac Merritt Singer. In ill health and suffering from several serious conditions, all the trials and excesses of his youth had come back to haunt his final days. He survived to see his beloved daughter married and went downhill rapidly afterwards.

At the Kilbowie it was not uncommon for three generations of the same family to work side by side, passing down the knowledge and skills to each new generation of workers.

Kilbowie would survive two World Wars but not the ever creeping expansion of throw-away plastic from the emerging economies. The last sewing machine came off the assembly line in June of 1980.

Said to be the largest clock tower in the world with four back-lit faces 26ft wide. The Singer clock was a sight to behold. Each bronze hand weight over a 2,000lbs.

The largest clock tower in the world, which once summoned the whole valley to work each day, was dismantled in front of the eyes of the remaining workforce. Many cried. Some of it was melted down and sold as souvenir ash trays and other bits of tourist tat.

The mechanics of the clock were scuttled away to America never to be seen again. Many of the workers, who had taken over from their fathers and their fathers before them, took bits of it to remember the business that had brought employment to generations of working families.

Late in 1980 the great factory finally closed its doors. It was abandoned, then demolished.

With the loss of the ship building coinciding with the closure of the Singer Kilbowie factory, Clydebank became a ghost town.

It was the end of an era, the end of a dream, and the end of the British sewing machine industry.

The End

The Rise And Fall Of Singer Manufacturing in Britain

End of Empire
By
Alex Askaroff
©

For all other publications by
Alex Askaroff
See Amazon

Isaac Singer
The First capitalist

Most of us know the name Singer but few are aware of his amazing life story, his rags to riches journey from a little runaway to one of the richest men of his age. The story of Isaac Merritt Singer will blow your mind, his wives and lovers his castles and palaces all built on the back of one of the greatest inventions of the 19th century. For the first time the most complete story of a forgotten giant is brought to you by Alex Askaroff.

No1 New Release. No1 Bestseller
Amazon certified.

*If this isn't the perfect book it's close to it!
I'm on my third run though already.
Love it, love it, love it.
F. Watson USA*

Elias Howe
The Man Who Changed The World
No1 New Release Amazon Oct 2019.

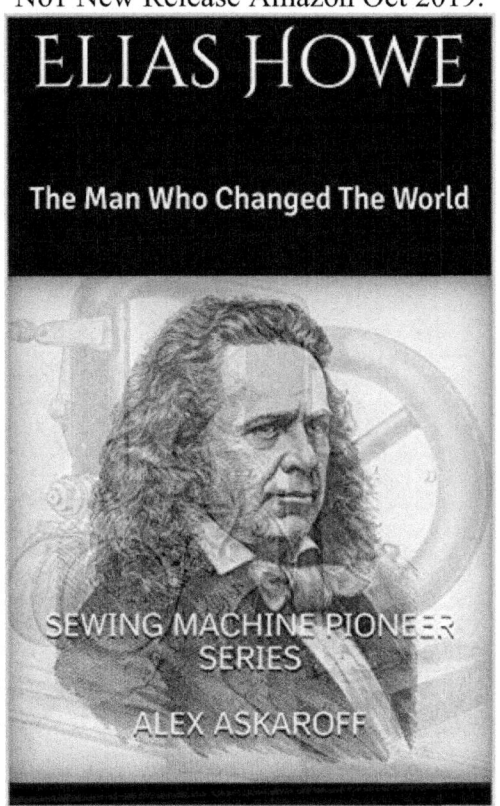

Anyone who uses a sewing machine today has one person to thank, Elias Howe. He was the young farmer with a weak body who figured it out. Elias's life was short and hard, from the largest court cases in legal history to his adventures in the American Civil War. He carved out a name that will live forever. Elias was 48 when he died. In that short time he really was the man who changed the world.